Short Notes on Applied and Advanced Inorganic Materials Chemistry

Short Notes on Applied and Advanced Inorganic Materials Chemistry

Dr. Ekambaram Sambandan Ph.D.

iUniverse, Inc.
New York Lincoln Shanghai

Short Notes on Applied and Advanced Inorganic Materials Chemistry

iUniverse books may be ordered through booksellers or by contacting:

iUniverse
2021 Pine Lake Road, Suite 100
Lincoln, NE 68512
www.iuniverse.com
1-800-Authors (1-800-288-4677)

ISBN-13: 978-0-595-39374-9 (pbk)
ISBN-13: 978-0-595-83770-0 (ebk)
ISBN-10: 0-595-39374-8 (pbk)
ISBN-10: 0-595-83770-0 (ebk)

Printed in the United States of America

DEDICATED TO
PHOSPHORTECH CORPORATION, USA.

CONTENTS

ACKNOWLEDGEMENTS

I thank all the members of PhosphorTech corporation, USA for their encouragement in writing the short notes. Also, I thank my professors, colleagues, relatives and well wishers for their support on my work. Finally, I acknowledge my wife, Kalyani and my daughter, Nandhini for their patience while writing the short notes during the sleeping hours in the night.

S. Ekambaram
PhosphorTech Corporation

SYNTHESES OF INORGANIC MATERIALS

1. Conventional Solid State Method[1-5]

Theoretical Principle:
   This method of synthesis of inorganic materials involves heating the corresponding component of oxides or carbonates in a stoichiometric ratio at a very high temperature in an inert container such as quartz or alumina or platinum. For example, oxide material synthesis requires temperature in the range 500-2000°C.
   This is a diffusion control reaction because of solid state reaction. Therefore, high temperature heating is required. Fig.1 illustrates a solid state reaction. Thus, reactants, A & B are ground well and the physical mixture is heated at a particular temperature. Then, repeated grinding and heating are performed to make sure all the reactants are converted into product.

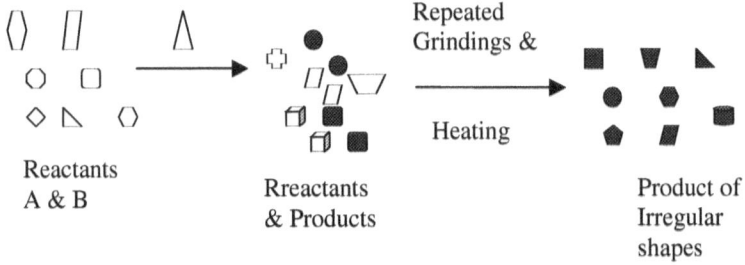

Fig. 1 Representation of a solid state reaction

   Once product is formed, it functions as a barrier for the reactants to react. Therefore, repeated grinding makes contact, for example, between two reactants so that they will react. Barrier concept is explained pictorially in the following reaction (Fig.2).

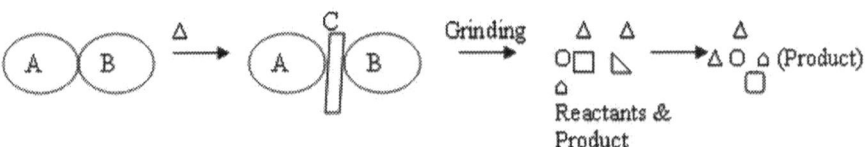

Where A and B are reactants whereas C is the product.

Fig.2 Application of repeated grinding and heating in a solid state reaction

The most advantage of solid state technique is that all the reactants are con-verted into solid product, free from pollution. A disadvantage of solid state technique is the particle size limitation. i.e. even after mechanical grinding the particle size is at most achieved to sub-micron. Therefore, wet-chemical meth-ods are useful in this regard. The works on wet-chemical methods are found in references [1-3].

Starting materials employed in the solid state technique are oxides or car-bonates or hydroxides. For an effective reaction starting materials should be particle size as low as possible. Therefore, carbonates as starting materials are found useful since they lead to formation of reactive oxides 'in situ' in this type reaction.

Experimental Procedure:

This is explained by considering solid state reaction between $SrCO_3$ and SiO_2 to get Sr_2SiO_4. Usually, pre-heat treatment well below the formation tem-perature is required as a warm-up treatment. For this reaction, the warm-up treatment is 600°C. Then, it is heated with temperature as high as 1300°C to get the required phase. Some time, post-treatment is needed for critical com-pounds such as high-temperature superconductor, $YBa_2Cu_3O_{6.9}$. In order to accelerate solid state reaction flux is added with reactants and finally the flux will be washed out using water.

In a typical experiment, 50g $SrCO_3$, 13.04g $BaCO_3$, 12.4g SiO_2, 1.73g EuF_3 and flux, 1.2g NH_4Cl are mixed well to get the nominal composition of $(Sr_{1.64}Ba_{0.32}Eu_{0.04})SiO_4$. This mixture is ball milled in presence of small amount of methanol at room temperature for 24 h. Then, this mixture is fired at 600°C for 2 h. Then, it is ground well and the final heat-treatment at 1300°C is carried out in the follow of mixture of H_2 and N_2 gases. Formation of host and presence of activator phosphor are confirmed by powder XRD and fluo-rescence spectra respectively. Emission spectrum obtained by exciting at 450 nm shows a broad band peaking at 560 nm.

2. Solid State Metathesis Method[6,7]

Theoretical Principle:

Dr. Kaner (USA) and Dr. Parkin (UK) independently reported the synthesis of a variety of materials by solid state metathesis (exchange) reactions. This method involves exchange exothermic reactions between reactive metal halide with alkali metal main group compound. The driving force for this type of reaction is the formation of thermodynamically stable salt such as AX where A=Na or K & X=Cl, Br, I. Prime importances of this type of reaction are

i.   preparation of anion solid solutions

ii.   Obtaining various particle sizes of products.

A serious limitation of this procedure is the requirement of anhydrous halides which require handling of reactants in dry box and storage in presence of inert atmosphere.

Experimental Procedure:

Technologically important materials such as superconductors, semiconductors, magnetic materials, intermetallics and so on are synthesized by this method. For example, the synthesis of ZrO_2 is explained here. ZrO_2 has three structures modification. One is monoclinic zirconia (room temperature form), the second is tetragonal zirconia and the third is cubic zirconia (high temperature forms). The phase formation depends upon temperature. However, t- or c- zirconia could be stabilized at room temperature by substituting Zr site of ZrO_2 by Y_2O_3, CeO_2, MgO or CaO. The reactants employed in this method for the synthesis of zirconia are anhydrous $ZrCl_4$ and Na_2O. Once igniting the starting mixture of $ZrCl_4$ and Na_2O, they become rapidly self-sustaining and can reach high temperature within a short period. The formation of zirconia is represented by the following equation.

$$\text{Ignition}$$
$$ZrCl_4 + 2Na_2O \rightarrow ZrO_2 + 4NaCl \qquad (1)$$

Thus, the reactant mixture consists of 1.0g $ZrCl_4$, 0.532g Na_2O. Na_2O_2 could also be used instead of Na_2O. These precursors are ground in an inert atmosphere. Then, it is ignited from a heated nichrome filament in an argon-filled glove box. After a few minutes, reaction gets over. Then, the product is washed with ethanol, followed by 2M HCl to dissolve an unreacted starting material. This reaction leads to formation of monoclinic zirconia. Tetragonal or cubic zirconia could be obtained by adding 9 atom% CaO or Y_2O3 or CeO_2. Now-a-days, this reaction is extended to synthesis complex oxides by Prof. J. Gapalakrishnan, India.

Both the cationic and anionic solid solutions could be synthesized by SSM. Thus, the SSM is extended to prepare (Mo, W)S_2 from $MoCl_5$ and WCl_3 precursor and $Mo(S,Se)_2$ is also prepared by a reaction among $MoCl_5$, Na_2S and Na_2Se. Therefore, I strongly believe that anion slid solution such as Zn(O,S) could also be prepared by the SSM. Because, ZnS is a photocatalyst for H_2

evolution from water and ZnO is good for O_2 evolution from water. Combination of these two catalyst might be a good photocatalyst for simultaneous evolution of H_2 and O_2 from water.

3. Aqueous Combustion Method[8,9]

Theoretical Principle:

This method of synthesis of oxides is new and hence, it is explained in detail.

Combustion process is an exothermic reaction, which occurs with evolution of light and heat. This is usually represented by a triangle (Fig.3) and each corner represents oxidizer, fuel and ignition.

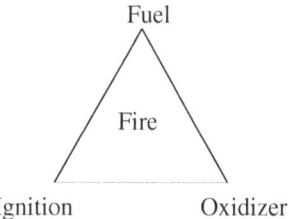

Fig.3 Explanation of Combustion

Therefore, it is important to ignite the mixture containing appropriate amounts of fuel and oxidizer. Combustion is simply expressed by a well-known reaction, burning of carbon in presence of oxygen.

$$C + O_2 \xrightarrow{\triangle} CO_2 + heat \qquad (2)$$

Here, carbon is reducer and O_2 is oxidizer. For the combustion synthesis of oxides, carbon is replaced by urea or glycine or hydrazine-based compounds. O_2 is replaced by metal nitrates, ammonium nitrate and ammonium perchlorate. Therefore, usually, oxidizer contains metal ion. But, sometime, fuel such as ammonium meta vanadate contains metal ion. Boric acid can be used as a neutral source for boron in the synthesis of borates.

Stoichiometric compositions of metal nitrates and fuel are calculated based upon propellant chemistry. Thus, heat of combustion is maximum for O/F ratio unity. Based on the concepts used in propellant chemistry, the elements

C, H, V, B, or any other metal are considered as reducing elements with valencies 4+, 1+, 5+, 3+ (or valence of the metal ion in that compound), respectively and oxygen is an oxidizer having the valence of 2-. The valence of nitrogen is taken as zero because of its conversion to molecular nitrogen during combustion.

The oxidizing and reducing valencies of metal nitrates and fuels used in the combustion synthesis of oxide phosphors are summarized in table 1.

Table1: Valences of metal nitrates and fuels

Compounds	Oxidizing or reducing valences
$M(NO_3)_2$	10-
$M(NO_3)_3$	15-
$M(NO_3)_4$	20-
NH_4NO_3	2-
Urea, CH_4N_2O	6+
Oxalyl Dihydrazide, ODH, $C_2H_6N_4O_2$	10+
Carbohydrazide, CH, $CH_6N_4O_2$	8+
3-methly-pyrazole-5-one, 3MP5O, $C_4H_6N_2O$	20+
Diformyl Hydrazine DFH, $C_2H_4N_2O_2$	8+
NH_4VO_3	3+
H_3BO_3	0

Calculation of molar ratio of reactants:

Al_2O_3 synthesis using $Al(NO_3)_3$ and urea is described in this section. Thus, for complete combustion of 2 moles of $Al(NO_3)_3$, 5 moles of urea is required.

$$2Al(NO_3)_3 + 5\,CH_4N_2O\,(Urea) \xrightarrow[500°C]{\triangle} Al_2O_3 + 5CO_2 + 8N_2 + 10H_2O \qquad (3)$$

15- (Oxidizing   6+ (reducing
    valence)   valence)

  2   5

  ___   ___

  30-   30+

Experimental Procedure:

Combustion method is yet another wet-chemical method which does not require calcinations and repeated heating. During combustion procedure lots of heat evolve. Such a high temperature leads to formation and crystallization of oxides.

Wet-chemical techniques are now available for simple oxide phosphor such as Eu^{3+} activated Y_2O_3 red phosphor. These techniques are defined as techniques which do not comprise of the normal mixing, calcinations and grinding operations. These wet-chemical methods dope rare-earth activators uniformly. But, calcination is required to get crystalline (required) phosphors. However, wet-chemical techniques are not available for the synthesis of complex oxide phosphor (green phosphor).

Tb^{3+} activated $(La,Ce)MgAl_{11}O_{19}$ green phosphor is obtained by rapidly heating an aqueous concentrated solution containing stoichiometric amounts of metal nitrates [$La(NO_3)_3$, $Ce(NO_3)_3$, $Tb(NO_3)_3$, $Mg(NO_3)_2$] and urea at 500°C. Thus, $M(NO_3)_2$:urea (1:1.66), and $M(NO_3)_3$:urea (1:2.5) redox compositions are used for the combustion synthesis of $(La,Ce)MgAl_{11}O_{19}:Tb^{3+}$. Theoretical equation assuming complete combustion can be written as follows:

$$1\text{-}xCe(NO_3)_3 + xTb(NO_3)_3 + Mg(NO_3)_2 + 11Al(NO_3)_3 + 31.67\ CH_4N_2O\ (urea) \xrightarrow{\triangle}$$
$$Ce_{1-x}Tb_xMgAl_{11}O_{19} + by\ products \qquad (4)$$

Powder XRD pattern of green phosphor reveals single phase crystalline in nature. This observation is notable because ceramic synthesis of green phosphor requires elevated temperature (>1400°C) and always contains Al_2O_3 as impurity.

Emission spectrum of combustion synthesized $LaMgAl_{11}O_{19}:Ce^{3+}$ shows a broad band at 340 nm. The emission of Ce^{3+} is due to $4f^65d \rightarrow 2F_j$ transition. Addition of Tb^{3+} in $LaMgAl_{11}O_{19}:Ce^{3+}$ results in emissions at 480 and 543 nm in addition to Ce^{3+} emission. The Tb^{3+} emissions which arise due to energy transfer from Ce^{3+}, are attributed to $^5D_3 \rightarrow {}^7F_5$ and $^5D_4 \rightarrow {}^7F_5$ transitions respectively. Thus, presence or doping of sensitizer (Ce^{3+}) and activator (Tb^{3+}) is confirmed by fluorescence spectra.

4. Hydrothermal Method

Theoretical Principle:

Theoretically speaking, hydro/solvo thermal technique could be employed for three major applications such as single-crystal growth, nano materials synthesis and recovery of metals from ores. This method involves heating the corresponding constituent in water/solvent to its above boiling point in a closed system with autogeneous pressure.

Now-a-days, this method is widely explored to synthesize novel open framework materials. Learn and lead methods direct to isolation of single crystals of open framework materials, when the reactants deviate from product stoichiometry. Once structure of open framework is solved and product composition is determined, stoichiometric reaction is carried out to get single phase polycrystalline powder.

Reactions are designed to carry out in water or solvent in presence of organic templates. Therefore, the product mostly contains water and hydroxyl group. Organic template directs the reaction to form an open framework structure with 1- or 2- or 3- dimensional pores.

Experimental Procedure:

Role of starting materials to obtain unique compounds is explained in this section. Usually, metal salts are used as reactants to get charge of metal ions in the product same as that of reactants. But, we are unique group to start with metal powder as starting material to synthesize mixed valence open framework materials [10].

For example, mixed valence titanium compounds are synthesized by starting with Ti metal powder. Otherwise, these compounds are not synthesized by any other methods. The reaction conditions and reactants molar ratio to synthesize mixed valence Ti compound are given in the following equation.

$$170°C, 1d, Autoclave$$
$$2Ti + 13.65H_3PO_4 + 500H_2O + 4.5 DAP \rightarrow$$
$$Ti^{III}Ti^{IV}(PO_4)(HPO_4)_2(H_2O).0.5DAP \qquad (5)$$

Where DAP = 1,3 diamino propane.

Other application for the synthesis of mixed valence compound is explained here. Titanium phosphate with mixed valence open framework material is obtained if diamino propane is replaced by ethylene diamine. Thus, the obtained product is $Ti^{III}Ti^{IV}(HPO_4)_4.C_2N_2H_9.H_2O$ [11]. This compound is found stable up to 600°C in air. At 600°C, enH is expelled out from the compound. For the charge compensation, Ti^{III} is getting oxidized to Ti^{IV} $(Ti_2(HPO_4)_4)$. Therefore, the structure is retained without destruction up to 600°C in air [11]. Thus, $Ti(HPO_4)_2$ is the first three dimensional titanium hydrogen phosphate reported recently.

5. Ion-exchange and Intercalation Methods

Theoretical Principle:

Ion-exchange reaction is a replacement of mostly cations by other cations such that the structure of reactant is retained in the product also. Intercalation is the reaction of interlayer ion with other reactant. Here also, the structure of reactant is retained in the product. These reactions are carried out at low-temperature and hence, called as chemie duce or soft reaction. Such compounds can't be synthesized by high-temperature reactions. Ion-exchange and intercalation reactions are represented in the following equation.

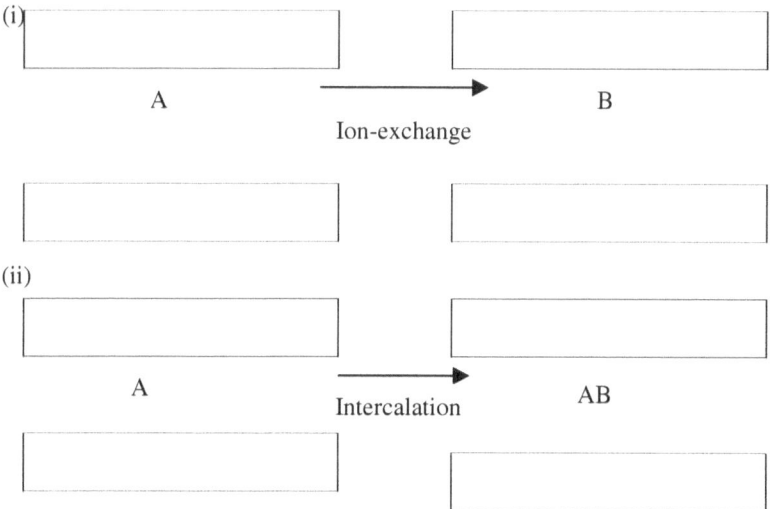

Fig. 4: Schematic diagram of ion-exchange and intercalation reactions.

Experimental Procedure:

(A) Ion-exchange method:

Cations present in the interlayer of hectorite (Na^+) are replaced by Cd^{2+} ion and hence, this type of reaction is called ion-exchange reaction.

$$\text{Ion-exchange with } Cd(CH_3COO)_2$$
$$\text{Hectorite/Na}^+ \xrightarrow[\text{RT}]{} \text{hectorite/Na}^+, Cd^{2+} \tag{6}$$

About 10 times excess $Cd(CH_3COO)_2$ with respect to hectorite weight is dissolved in water. To this clear solution hectorite is added and the solution is allowed for stirring at room temperature for 24 h. Then, this solution is filtered off to get white powder of Cd^{2+} ion-exchanged hectorite (hectorite/Na^+,Cd^{2+}). After ion-exchange reaction, the hectorite obtained is a neutral compound 'and therefore, some of Na^+ ions are replaced by Cd^{2+} ions. This colorless powder is used for intercalation reaction in the following part.

(B) Intercalation Reaction

In this section, intercalation reaction is described. Intercalation reaction between hectorite/Na^+,Cd^{2+} dispersed in water and H_2S gas (method 1) results in expulsion of CdS from the interlayer of hectorite as represented below.

$$\text{Hectorite/Na}^+, Cd^{2+} \text{ (aqueous)} + H_2S \text{ (gas)} \rightarrow \text{hectorite/Na}^+ + CdS \tag{7}$$

CdS has been incorporated successfully into the inter layer of hectorite by the intercalation between hectorite/Na^+, Cd^{2+} (solid) and H_2S gas (method 2) at room temperature [12]. This is an important reaction since reactants are solid and gas phases. Incorporation of CdS in the interlayer of hectorite was confirmed by powder XRD patterns. Hectorite/Na^+, CdS prepared by method 2 does not show reflections corresponding to CdS whereas this material prepared by method 1 shows the CdS reflections in the powder XRD pattern. Incorporation of CdS in the interlayer is further supported by UV-Vis absorption.

6. Reverse Micelle or Micro emulsion Method [13]

Theoretical Principle:

When surfactants in hydrocarbon solvents are mixed with a little quantity of water, inverse micelle or reverse micelle forms. It is made up of spherical shape where in outer surface is hydrocarbon end of surfactants and inner surface is ionic end of surfactants. Thus, inner surface acts as ionic vessels for ionic reaction. The formation of reverse micelle or inverse micelle is represented below (Fig.5).

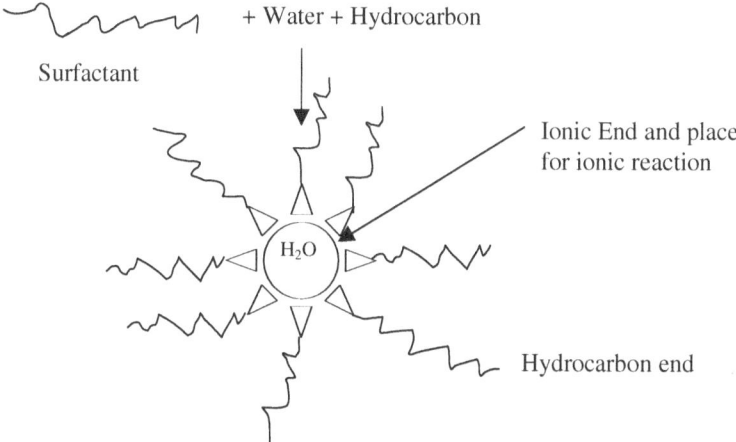

Fig.5: Water microemulsion formation of surfactants in hydrocarbon medium.

Experimental Procedure:

Formation of nanomaterials with various sizes is understood here. The sizes of nanomaterials may be varied by changing the ratio of water/hydrocarbon solvent.

For example, Cd^{2+} ions are dissolved in reverse micelles. At this stage, Cd^{2+} ions occupy in inner sphere of surfactants i.e. in water droplets. Then, H_2S gas is passed through the reverse micelle as a source of S. Since H_2S is an ionic gas, it is getting touch with Cd^{2+} ions. CdS is formed in water droplets. Thus, CdS of nanometer size could be prepared by reverse micelle method. The following diagram with an equation describes the formation of CdS in the reverse micelles (Fig.6).

Micro-emulsion + Cd^{2+} + S^{2-}

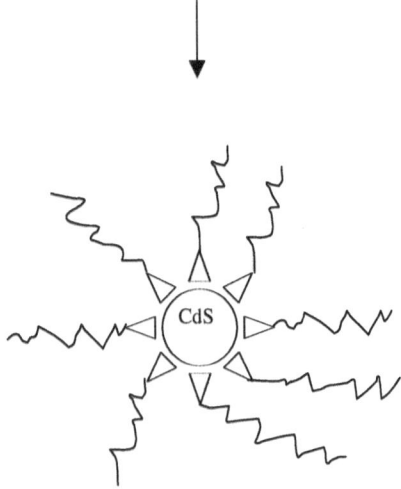

Fig.6 CdS formation in microemulsion

7. Precursor Method

Theoretical Principle:

Importance of precursor method for the synthesis of inorganic materials is clearly understood by considering the formation of cation solid solution. Formation of cation solid solution is evidenced from similar powder XRD patterns of individual cation compound. When cation solid solution is rapidly heated above its decomposition temperature, fine particles with better properties of inorganic materials are obtained. If the ligands used to get cation solid solution contains large combustible group, then, large no. of gases evolved during decomposition so that the final product of inorganic materials exhibits better properties than the same materials synthesized by conventional solid state technique. An example of ligand for the synthesis of cation solid solution is hydrazine carboxylate, $N_3H_3COO^-$. It is a fuel rich ligand and hence, it takes atmospheric oxygen during decomposition to get metal oxides.

Experimental Procedure:

Precursor method of synthesis of inorganic materials is explained using Eu^{3+} doped yttrium hydrazine carboxylate compound. Eu^{3+} and Y^{3+} form compounds with hydrazine carboxylate ligand $(N_2H_3COON_2H_5)$. These compounds are $Y(N_2H_3COO)_3 \cdot 3H_2O$ and $Eu(N_2H_3COO)_3 \cdot 3H_2O$. Powder XRD patterns of these compounds are very similar to each other and hence, these compounds form solid solution when $EuCl_3$ and YCl_3 are treated with $N_2H_3COON_2H_5$ at room temperature [8].

$$\overset{RT}{0.99YCl_3 + 0.01EuCl_3 + N_2H_3COON_2H_5 \text{ (excess)} \rightarrow (Y_{0.99}Eu_{0.01})(N_2H_3COO)_3 \text{ (8)}}$$
$$\text{Cation solid solution}$$

This cation solid solution is decomposed in air in the temperature range 300-1000°C to get red phosphor. Also, presence of oxidizer such as NH_4NO_3 or NH_4ClO_4 reduces the decomposition temperature of the cation solid solution to 300°C.

$$2(Y_{0.99}Eu_{0.01})(N_2H_3COO)_3 \overset{600°C}{\rightarrow} (Y_{0.99}Eu_{0.01})_2O_3 \qquad (9)$$
$$\overset{300°C}{\underset{NH_3NO_3 \text{ Or } NH_4ClO_4}{\rightarrow}} (Y_{0.99}Eu_{0.01})_2O_3 \qquad (10)$$

8. Co-precipitation Method

Theoretical Principle:

This method involves simultaneously precipitating two or more metal salts to form intimate mixtures or solid solutions (precursor). Since water is used as solvent medium it is also called a wet-chemical method. The precursor is obtained either by adding a precipitating agent or by evaporating the solvent. The following equation represents the principle of co-precipitation as intimate mixture of two metal hydroxides.

$$\overset{\text{Addition of } OH^-}{MCl_2 + 2M'Cl_3 + H_2O \rightarrow M(OH)_2 + 2M'(OH)_3 \text{ (precursor)}} \qquad (11)$$

$$M(OH)_2 + 2M'(OH)_3 \xrightarrow{\triangle} MM'_2O_4 \qquad (12)$$

Since precursor is an intimate mixture of two metal ions, formation temperature of MM'_2O_4 is lowered when compare to that of solid state reaction between $M(OH)_2$ and $M'(OH)_3$.

Experimental Procedure:
Formation of $ZnFe_2O_4$ is explained in this section. Zn-oxalate and Fe-oxalate are dissolved in water. Then, water is evaporated to get precursor of oxalates of zinc and iron. Here, the precursor is a solid solution of zinc and iron oxalates. Then, this solid solution is heated to get powder of $ZnFe_2O_4$ compound. The following equation represents the formation of $ZnFe_2O_4$ powder.

$$Zn(C_2O_4) + Fe_2(C_2O_4)_3 + H_2O \xrightarrow{\text{Evaporation of water}} ZnFe_2(C_2O4)_4$$
$$\xrightarrow{\triangle} ZnFe_2O_4 + 4CO + 4CO_2. \qquad (13)$$

The byproducts are gases, obtained solid product by this method is only $ZnFe_2O_4$. A disadvantage of this method is that precipitation of two or more cations, simultaneously, is critical and requires extreme care.

9. Sol-gel Method

Theoretical Principle:
The sol-gel method is used extensively for the synthesis of silicates and some oxides. It involves three steps and these are

1.   Formation of sol
2.   Conversion of sol into gel
3.   Decomposition of gel

The first step is obtained when molecules combine to yield small particles. Therefore, sol is the dispersion of small particles of oligomers in solvent. The second step is the evaporation of solvent to get extended polymeric gel. This step is crucial to obtain mixed oxides/silicates. The final step is the calcinations of gel to get corresponding inorganic materials. The following equations represent about the formation of silicates.

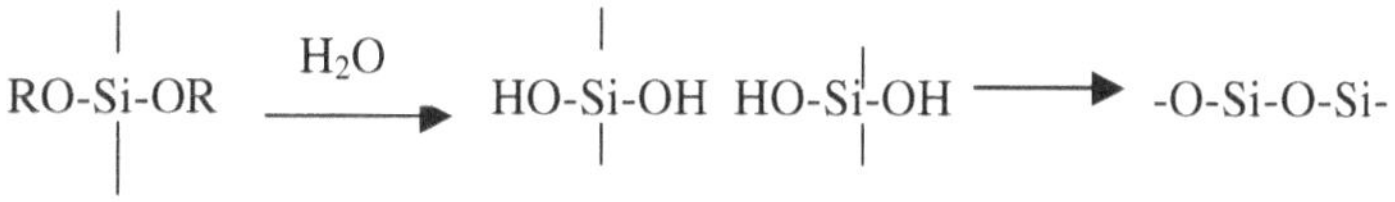

$$\text{RO-Si-OR} \xrightarrow{\text{H}_2\text{O}} \text{HO-Si-OH HO-Si-OH} \longrightarrow \text{-O-Si-O-Si-} \qquad (14)$$

Experimental Procedure:

Formation of M-O-M' bond is described using sol-gel method in this section. For this (formation of M-O-M' bond) alkoxide precursor is required. Thus, alkoxide reacts with water to form M-OH bond, which is followed by condensation to yield M-O-M bonding. The typical example of metal alkoxide is $Ti(O^iPr)_4$ [Titanium iso propoxide] and synthesis of TiO_2 by sol-gel method is represented below.

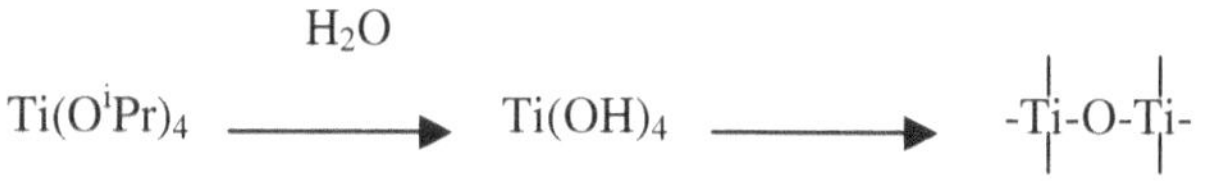

$$Ti(O^iPr)_4 \xrightarrow{\text{H}_2\text{O}} Ti(OH)_4 \longrightarrow \text{-Ti-O-Ti-} \qquad (15)$$

Relevant References:

1. CNR Rao, Chemical synthesis of solid inorganic materials, *Mater. Sci. Eng.* **B18**, pp.1-21, 1993.

2. CNR Rao, Chemical Approaches to the synthesis inorganic materials, New Delhi, Wiley Eastern Limited, 1994.

3. D. Segal, Chemical synthesis of ceramic materials, *J. Mater. Chem.*, 7, pp.1297-1305,1997.

4. R. Nagarajan and CNR Rao, Structure and superconducting properties of Ga-substituted $YBa_2Cu_3O_{7-\delta}$ and $YBaSrCu_3O_{7-\delta}$ systems, *J. Mater. Chem.*, 3, pp. 969-973, 1993.

5. J.S. Yoo, S.H. Kim, Y.T. Yoo, G.Y. Hong, K.P. Kim, J. Rowland and P.H. Holloway, Control of spectral properties of strontium-alkaline earth-silicate-europium phosphors for LED applications, *J. Electrochem. Soc.*, 152, No. 5, pp.G382-G385, 2005.

6. E.G. Gillan and R.B. Kaner, Rapid, energetic metathesis routes to crystalline metastable phases of zirconium and hafnium dioxide, *J. Mater. Chem.*, 11, pp. 1951-1956, 2001.

7. T. Sivakumar, J. Gopalakrishnan, Transformation of Dion-Jacobson phase to aurivilies phase: synthesis of $(PBBiO_2)MNb_2O7$ (M = La, Bi), *Mater. Res. Bull.*, 40, pp.39-45, 2005.

8. S. Ekambaram, K.C. Patil and M. Maaza., Synthesis of lamp phosphors:facile combustion approach, *J. Alloys and Compounds*, 393, pp. 81-92, 2005.

9. K.C. Patil, S.T. Aruna and S. Ekambaram, Combustion synthesis, *Curr. Opin. Solid State mater. Sci.*, 2, pp. 158-165, 1997.

10. S. Ekambaram and S.C. Sevov, Organically templated mixed valent Ti^{III}/Ti^{IV} phosphate with an octahedral-tetrahedral open framework, *Angew. Chem. Int. Ed.*, 38, pp. 372-375, 1999.

11. S. Ekambaram, C. Serre, G. Ferey and S.C. Sevov, Hydrothermal synthesis and characterization of an ethylene diamine templated mixed valence titanium phosphate, *Chem. Mater.*, 12, pp. 444-449, 2000.

12. S. Ekambaram et. al., unpublished work.

13. Ulrich Schubert and Nicola Husing, Inorganic materials synthesis, Wiley-VCH, 2004.

POWDER XRD FOR THE CHARACTERIZATION OF MAINLY INORGANIC MATERIALS

Powder X-ray diffraction is a valuable technique for solid state and inorganic materials scientists. Even though, it has long history, now-a-days it is a main and fast technique for the characterization of mainly inorganic materials. Before going to describe the importance of powder XRD let me very briefly explain the derivation of Bragg's equation. This equation is simple but it explains depth of structure of polycrystalline and single crystal materials. All scientists involving in solid state and inorganic materials synthesis rely on this equation.

Bragg's Equation:

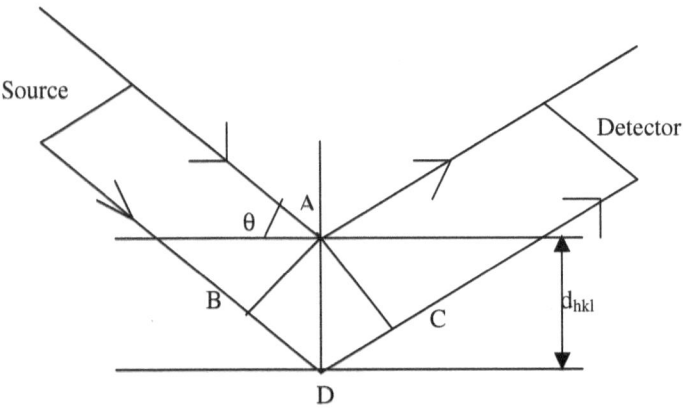

Fig.7 Diffraction of X-rays by a crystalline material

Wavelength of X-rays is in the order of difference between lattice planes of crystals. Therefore, crystals behave as grating for X-rays. Hence, X-rays are used to solve structure of crystals.

Let us consider that atoms or ions as points arranged in a regular fashion in planes. When the polycrystalline powders are exposed to X-ray with an unique wavelength, λ it is diffracted by array of ions (Fig.7). If diffracted waves are in-phase, amplitude of them is doubled from the incident wavelength. If diffracted waves are out of phase, then amplitude of them is zero. Therefore, detector sees diffracted waves only if they are in-phase.

Let us assume the angle of incident wavelength (A and D points) of X-ray at the plane of atoms in a crystal is θ.

The path difference between two lattice planes depends upon wavelength of X-ray and it is equal to $n\lambda$ (16)

BD and DC are the extra distances of X-ray traveled for the diffraction at D from diffraction at A. Therefore, path difference = BD + DC = $2d_{hkl} Sin\theta$ (17)

Comparing equations (17) and (18), we get
$n\lambda = 2d_{hkl} Sin\theta$ (18)
This equation is called Bragg's equation.

Rewriting the equation (19),
$\lambda = 2d_{hkl} Sin\theta/n$ (19)
n = 1 for a crystalline system. Then, equation (20) becomes

$\lambda = 2d_{hkl} Sin\theta$ (20)
<u>Note: Reason for n = 1</u>

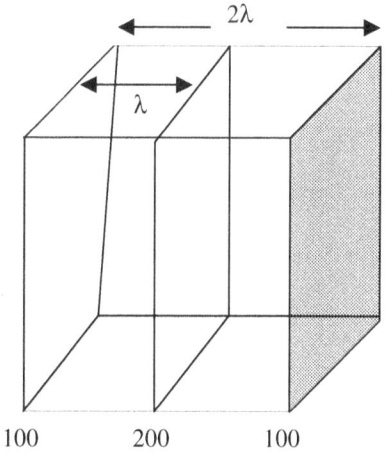

Fig.8: Diagram of 100 and 200 planes in an unit cell

The Bragg equation is $d_{hkl} = n\lambda/2sin\theta$ (21)

For second order diffraction from 100 plane, the Bragg equation becomes

$d_{100} = 2\lambda/2sin\theta = \lambda/sin\theta$ (22)

For the first order diffraction from 200 plane, the Bragg equation becomes

$d_{200} = \lambda/2sin\theta$ or

$$d_{100}/2 = \lambda/2\sin\theta = \lambda/\sin\theta \tag{23}$$

Equation (23) = Equation (24).

STRUCTURES OF A FEW INORGANIC MATERIALS

(A)   Structure of NaCl

NaCl is an ionic compound and the structure of NaCl is a simple cubic with Cl^- ion occupies at the corner and face centered of a simple cube whereas Na^+ ion occupies at the edges of cubic system (Fig.9).

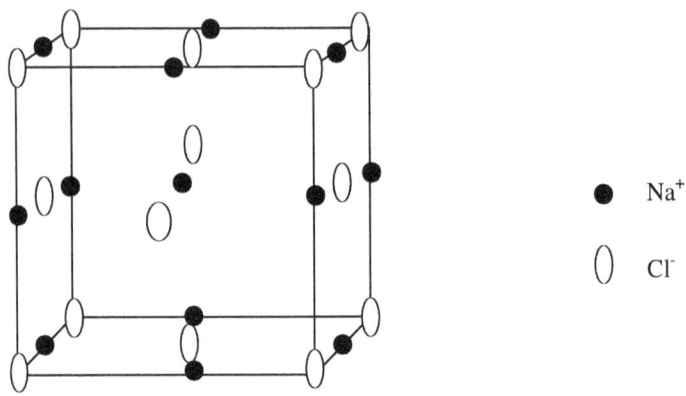

$\bullet$ Na^+

$\bigcirc$ Cl^-

Fig.9: Structure of oxides belonging to NaCl

The structure belonging to NaCl should be ionic. The coordination number is six for each Na^+ and Cl^-. The ionic radius ratio, r_{cat}/r_{anion} is between 0.414 and 0.732. For NaCl structure, ionic size difference is large and for CsCl structure, ionic size difference is not a large. Compounds or oxides deviating from ionic radius ratio may lead to non-stoichiometric for adopting the NaCl structure.

(B)   Structure of CsCl

It is also cubic structure and Cl^- ion occupies each corner and the large ion, Cs is fixed at the centre of cube (Fig.10). Thus, coordination number of Cs is 8 and for Cl ion it is six. It is also ionic solid. Because Cs is a bigger ion, it does not occupy at the edge of a cube. Radius ratio of CsCl structure is $1 > r_+/r_-$ >0.732. Transition metal ions are smaller in size and therefore, they do not form CsCl structure.

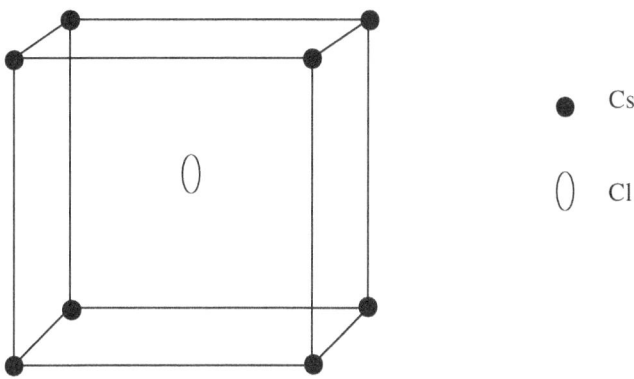

Fig.10: Structure related to CsCl

(C) Structure of Perovskite

Perovskite structure was named after the discovery of $CaTiO_3$ by the Perovski scientist. This structure is represented by ABO_3 (Fig.11). In this structure, the size of A-ion is larger than that of B-ion. Variety of compounds exhibiting this structure finds plethora applications for example, high-temperature superconductors.

The perovskite structure is shown below.

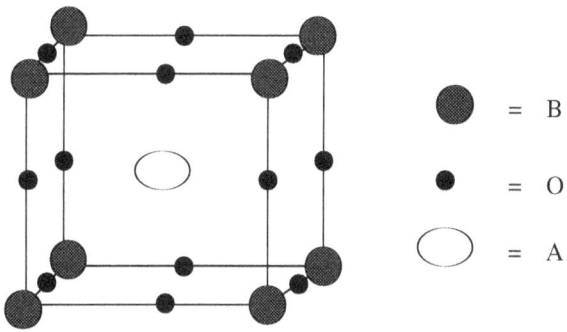

Fig.11: Structure of perovskite

The coordination geomentry of B cation is octahedral whereas that of A cation is 12. Examples for different perovskite compounds are $LaAlO_3$, $SrSiO_3$ and $NaWO_3$.

SOME RECENT DEVELOPMENTS

Photocatalysts for Water Splitting

I.   Introduction:

There are two vital applications of semiconductors in chemistry field. One is the conversion of photon energy into electrical energy or chemical energy. Another application is the water purification. Photosplitting of water into H_2 and O_2 is an up-hill reaction and it is a reversible reaction. Therefore, $\triangle G$ is greater than zero. Whereas detoxification of organic compounds in aqueous medium is down-hill reaction and it is an irreversible reaction. Therefore, $\triangle G$ is less than zero.

I.A. Principle of Photosplitting of water using semiconductors.

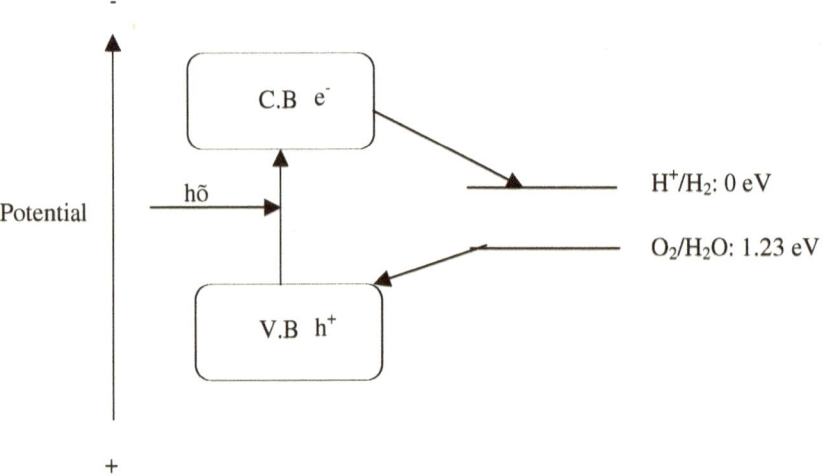

Fig.12: Use of semiconductor in photo-splitting of water

Level of bottom of conduction band should be more negative than that of redox potential of water into hydrogen and level of upper of valence band should be more positive than that of redox potential of water into oxygen for a semiconductor to be a photosplitting of water into H_2 and O_2 (Fig.12). The potential difference between these two levels of H^+ into H_2 and H_2O into O_2 is 1.23 eV. Therefore, semiconductors with band gap greater than 1.23 eV is required. When a semiconductor is exposed to wavelength greater than that of band gap, charge carriers are produced. Electrons are promoted to the CB whereas holes remain in the VB. These charge carriers are responsible for photosplitting of water into H_2 and O_2 as shown in the Fig.12. Sometime sacrificial reagents are used to get either H_2 or O_2. In a real system, there are a few reactions taking place. Once, charge carriers are formed in semiconductor particles, charge carriers should migrate to the surface of particles (charge separation) before they recombine. Then, the charge carriers involve in the photosplitting of water.

II.   Photosplitting of water:

Based upon photon energy, there are two available methods. One is under UV and the another one is under visible irradiations.

II.A. Photosplitting of water into H_2 under UV irradiation [14]

This is achieved using semiconductor nanocomposites viz. $H_4Nb_6O_{17}$/Pt, TiO_2, in presence of CH_3OH as a sacrificial reagent. $H_4Nb_6O_{17}$/Pt,TiO_2 could be synthesized from $K_4Nb_6O_{17}$ by soft-chemical routes. Photoproduction of H_2 from H_2O using $K_4Nb_6O_{17}$ and its modifications under UV-irradiation in presence of methanol is shown below (Fig.13).

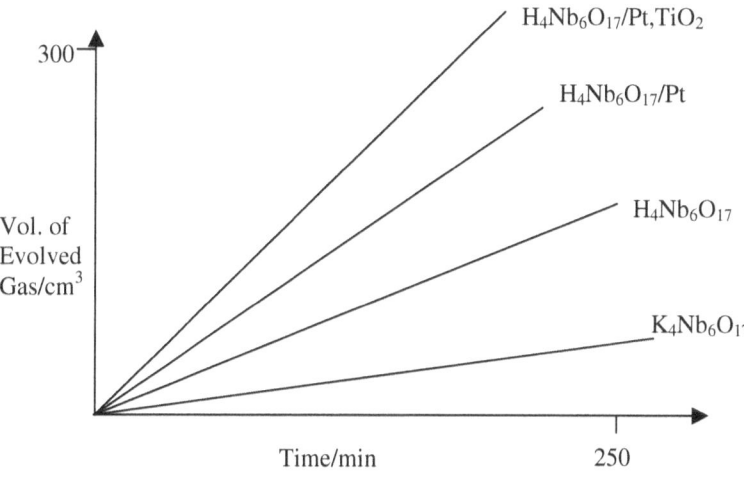

Fig.13: Rate of H_2 evolution

As seen in the Fig. 13, the highest catalytic activity is observed for $H_4Nb_6O_{17}$/Pt,TiO_2. This may be due to charge carriers separation between two semiconductors (TiO_2 and $H_4Nb_6O_{17}$) and hence, the life time of charge carriers is increased. Therefore, there is an enhancement in H_2 production in the case of $H_4Nb_6O_{17}$/Pt,TiO_2.

II.B. Photosplitting of H_2O into O_2 under visible irradiation [15]

One way of obtaining photocatalysts for this application is the doping of transition metal ion in semiconductors. Let consider the cobalt doped ZnO. It shows visible absorption. Therefore, it is expected to exhibit photosplitting of water in the visible irradiation. Thus, $Zn_{0.9}Co_{0.1}O$ having band gap of 2.7 eV (green color) shows 18.5 µmol/h O_2 evolution in presence of $AgNO_3$.

Relevant Reference:

14. S. Ekambaram, M.Yanagisawa, S. Uchida, Y. Fujishiro and T. Sato, "Synthesis and photocatalytic property of hectorite/(Pt. TiO_2) and $H_4Nb_6O_{17}$/(Pt, TiO_2) nanocomposites", *Mol. Cryst. And Liq. Cryst.,* 2000, 341, 213-218.

15. S. Ekambaram et.al, unpublished results

Diluted Magnetic Semiconductors[16,17]

Diluted magnetic semiconductors (DMS), also referred to as semimagnetic semiconductors, currently receive great attention owing to their potential applications in spintronics, magnetic switches and magnetic recordings. To obtain DMS, lattice ions in semiconductors are partly substituted by magnetic elements, such as transition metal or rare-earth (Fig.14).

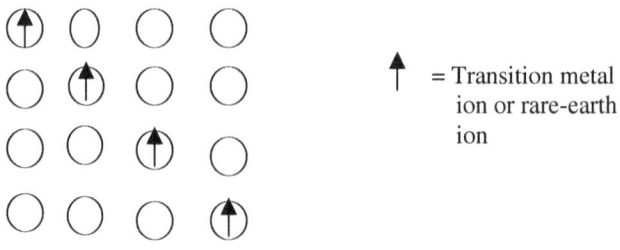

↑ = Transition metal ion or rare-earth ion

Fig.14: Representation of DMS.

In DMS quantum dots (characteristic size <100 nm), spins are used as information carriers and therefore, in addition to charge, electron spin is controlled to introduce new functionality in semiconductor devices. The most extensively studied DMS are $A^{II}_{1-x}Mn_xB^{VI}$ alloys where A = Cd, Zn, Hg etc. and B = S, Se, Te. Some examples are $Cd_{1-x}Mn_xSe$, $Zn_{1-x}Mn_xS$ and $Hg_{1-x}Mn_xTe$. More recently, attention has been focused primarily on oxides which provide superior properties, including stability. In this context, bulk and nano DMS have been realized in $Zn_{1-x}Mn_xO$, $Zn_{1-x}Co_xO$ and $Zn_{1-x}Ni_xO$. The ferromagnetic properties of DMS stem from the interaction of lattice, spin and electronic degrees of freedom between magnetic ions and semiconducting bulk/nano materials. Prediction of room-temperature ferromagnetism on transition metal ions doped ZnO by computations was verified by the preparation of thin films of Co^{2+}, Fe^{2+}, Mn^{2+} and V^{2+} doped ZnO.

Despite the recent progress in studies of DMS, some challenges remain for their production. First, it is difficult to reach uniform distribution of dopants in semiconductors for the required small concentrations. Another problem is segregation of transition metal oxides from solid solution due to high temperature and long processing time in the available methods, such as laser ablation, inverse micelle, solid state techniques, etc. Further, the properties of DMS depend on preparative methods, which are not universal for all DMS. Because of these issues, contradicting results on room temperature ferromagnetism of these materials have been reported in the literature. In addition, the available methods allow synthesis of only simple compound with two elements, such as oxides doped with transition metal ions. However, it is advantages to also produced mixed oxides (three or more elements) as hosts for transition metal and rare-earth ions. Finally, it is more difficult to synthesize DMS quantum dots as compared to bulk DMS. Thus, it is important to develop a novel synthetic method for uniform doping of transition metal and/or rare-earth ions in semiconductors, which allows production of DMS quantum dots based on both simple and mixed oxides.

The aqueous combustion method is explored to synthesize transition metal ions doped ZnO. However, magnetic measurement reveals absence of ferromagnetism even uniform doping of transition metal ions in ZnO. Color of $ZnO:Co^{2+}$ is blue, which exhibits ferromagnetism. But, the aqueous combustion method yields green color of $ZnO:Co^{2+}$ and this compound exhibits absence of ferromagnetism. Therefore, in addition to uniform doping of transition metal ions in semiconductor, there do have other factors that contribute ferromagnetism.

Relevant References:

16. CNR. Rao and F.L. Deepak, Absence of ferromagnetism in Mn- and Co- doped ZnO, *J. Mater. Chem.*, **15** pp. 573-578, 2005.

17. K.R. Kittilstved and D.R. Gamelin, Activation of High-Tc ferromagnetism in Mn2+ doped ZnO using amines, *J. Am. Chem. Soc.*, **127**, pp. 5292-5293, 2005.

Inorganic Phosphors for Solid State Lighting [18,20]

Phosphors find various applications ranging from fluorescent lamp to immunoassay. Among them, fluorescent lamp is largely used and hence, lot of efforts is focused on it. Now, there is an immense research activity in academia and industries in inorganic phosphors after the discovery of blue or near UV LED. The combination of blue LED and yellow phosphor leads to replace the Hg discharge fluorescent lamp. Thus, toxic mercury could be successfully replaced now-a-days. Phosphors in the blue LED convert part of blue emission/near UV of LED into green and red regions, and the combination of these colors results in white light. Nakamura patented yttrium aluminum garnet (YAG:Ce^{3+}) doped with Ce^{3+}, which is the best phosphor known today for obtaining white light from blue LED. Here, the part of blue light is transferred to yellow. Thus, the combination of blue and yellow leads to white light. The following figure illustrates the function of YAG doped with Ce^{3+} (Fig.15).

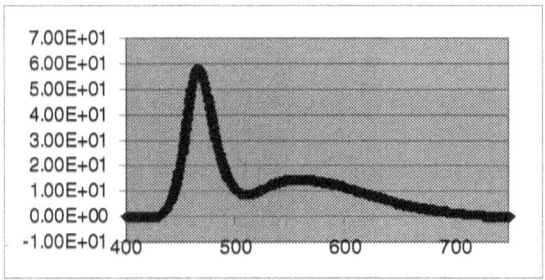

Fig.15: emission spectrum of YAG:Ce^{3+} under blue LED

The coordination number of Ce^{3+} in YAG is eight. Therefore, 8-coordinated ion in host materials is required for the development of non-YAG materials. Therefore, other host for Ce^{3+} can be thought of YVO_4 where coordination of yttrium in YVO_4 is eight. The general formual of pyrochlore structure is $A_2B_2O_7$. In this compound, A ion occupies eight coordination number, B ion occupies at octahedral coordination (Fig.16) and O ion occupies at tetrahedral coordination.

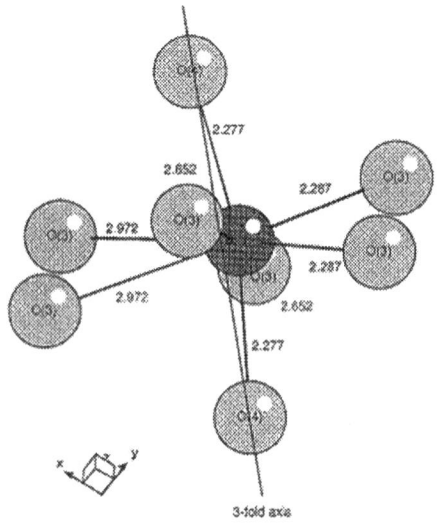

Fig.16: Coordination number view for A^{3+} ion

Thus, this compound is suitable for substituting A ion by Ce^{3+}. Typical examples for pyrochlore structure are $Ln_2Ti_2O_7$, $Ln_2Zr_2O_7$, $Ln_2Sn_2O_7$, $Ca_2Sb_2O_7$ etc. These compounds are potential candidates for Ce^{3+} doping and exhibiting yellow emission under blue LED.

There are two known activators, Ce^{3+} and Eu^{2+} for the LED solid state lighting. These activators in a particular host exhibit required emissions. It is usually observed that Ce^{3+} doped phosphor shows emission about 100-150 nm lower than that of Eu^{2+} emission in the same host. Therefore, it is a thumb rule in the solid state lighting to predict the emission wavelengths for Ce^{3+} and Eu^{2+} activators.

Relevant References:

18. T. Justel, H. Nikol and C. Ronda, New developments in the field of luminescent materials for lighting and displays, *Angew. Chem. Int. Ed.*, 37, pp. 3084-3103, 1998.

19. S. Neeraj, N. Kijima and A.K. Cheetham, Novel red phosphors for solid state lighting the system $Bi_xLn_{1-x}VO4:Eu^{3+}/Sm^{3+}$ (Ln = Y & Gd), *Solid State Commun.*, **131**, pp. 65-69, 2004.

20. S. Neeraj, N. Kijima and A.K. Cheetham, Novel red phosphors for solid-state lighting; the system $NaM(WO_4)_{2-x}(MoO_4)_x:Eu^{3+}$ (M =Gd, Y, Bi), *Chem. Phys. Lett.*, **387**, pp. 2-6, 2004.

Open Framework Materials [21]

There has now been a tremendous research work in the area of organically templated open framework materials. The open framework materials differ from thermodynamically stable condensed materials, and ring structures that usually observed in open framework materials are absent in the condensed materials. These open framework materials resemble well-developed zeoletic materials. The zeoletic materials are usually aluminosilicates. The differences between zeolites and organically templated open framework materials are summarized in the following table.2.

Table 2: comparison of properties of zeolites with open framework materials

Zeolites	Open Framework Materials
Alumino Silicates	Transition metals also present
Acid Catalysts	Could be a redox catalysts
Tetrahedral frameworks	Can be 5- or 6- coordination in addition to tetrahedral
Thermally stables	Not with all the compounds
Diamagnetic materials	Can be magnetic materials
Non-luminescent	Can be luminescent materials
Three-dimensional pores	1- or 2- or 3-dimensional pores

The stability of organically templated open framework materials is questionable. Templated organic molecules form extensive hydrogen bonding with framework oxide or hydroxyl group. Therefore, calcination destroys the hydrogen bonding, and open framework materials lead to the formation of stable condensed materials. For example, except a few compounds in organically templated metal phosphates, all the compounds loose the thermal stability.

Combination of open framework materials with semiconducting property may lead to removal of organic molecules occluded in the pores of open framework by photochemically without destroying the structures. Thus, protonated form of the compound is obtained and this type of conversion is represented below (Fig. 17).

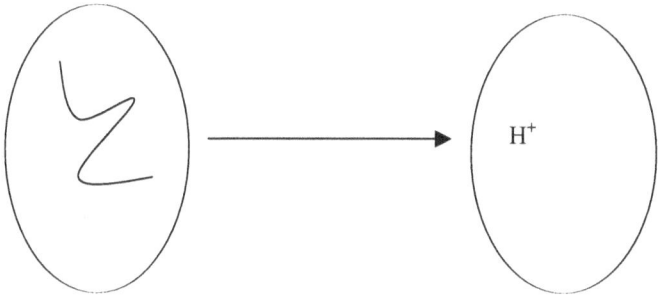

Fig.17: Formation of protonated open framework material

The protonated form could be

1. better photocatalysts for photosplittings of water
2. proton conductors.

Relevant Reference:

21. A.K. Cheetham, G. Ferey and T. Loiseau, Open-framework Inorganic Materials, Angew. Chem. Int. Ed., 1999, **38**,3269.

Three Way Catalysts [22,23]

In a chemical reaction, conversion of reactants into products takes place through a energy barrier (transition state) and the height of the barrier is represented by $e^{-Ea/RT}$ where Ea is called activation energy, T is the temperature and R is the gas constant. It is represented by the following diagram (Fig.18).

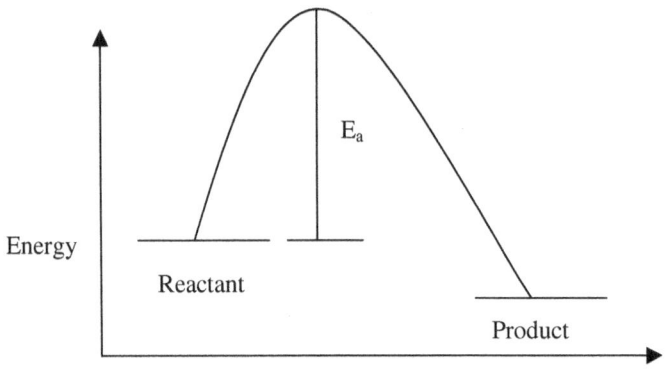

Fig.18: Representation of a reaction using activation energy.

Reaction can be accelerated by two ways. One way is by increasing temperature. Another way is by using catalyst, which actually decreases Ea by lowering the transition state. This is explained in the diagram below (Fig.19). Ea' is lower than that of Ea since catalyst reduces activation energy.

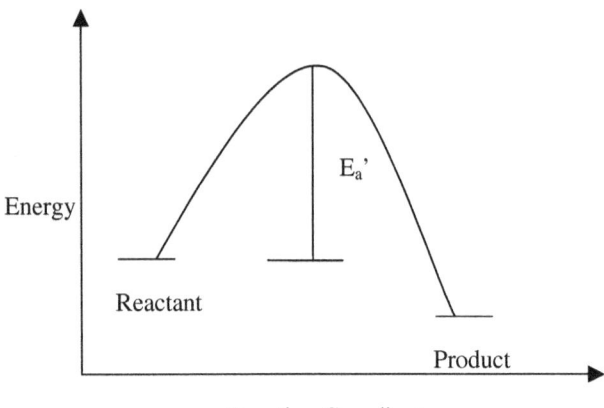

Fig. 19: Effect of catalyst on a reaction

In three way catalysis, the following reactions take place in an automobile exhaust.

Oxidation:

$$2CO + O_2 \rightarrow CO_2 \tag{24}$$

$$HC + O_2 \rightarrow CO_2 + H_2O \tag{25}$$

Reduction:

$$2CO + 2NO \rightarrow 2CO_2 + N_2 \tag{26}$$

$$HC + NO \rightarrow CO_2 + H_2O + N_2 \tag{27}$$

where HC=unburned hydrocarbon

In order to achieve the oxidation and reduction reactions, catalyst should be bi-functional, i.e. catalyst oxidizes CO and unburned hydrocarbons into CO_2 and H_2O and reduces NO into N_2 and H_2O. Conventional catalyst is Pd dispersed Al_2O_3 or CeO_2-ZrO_2 composite. The main drawback of the conventional catalyst is that particle growth occurs at such a high temperature in the automobile exhaust. Therefore, activity is decreased due to lowering in surface area. Of late, Pd-perovskite was found to be self-regeneration for automotive emissions control. The intelligent catalyst is $LaFe_{0.57}Co_{0.38}Pd_{0.05}O_3$. The mechanism of the intelligent catalyst in automotive emissions control is as follows. During oxidation and reduction atmospheres structural changes takes place in the intelligent catalyst. Thus, during reduction, Pd is getting released from the intelligent catalyst and Pd gets into the perovskite catalyst in oxidation atmosphere. At high temperature in automobile exhaust, perovskite catalyst undergoes such changes and thus, particle size is not affected and maintains catalytic activity unlike observed particle growth in the conventional catalyst. Therefore, it is highly desirable to have a catalyst that accommodates Pd in its structure during oxidation process.

Relevant References:

22. Y. Nishihata, J. Mizuki, T. Akao, H. Tanaka, M. Uenishi, M. Kimura, T. Okamoto and N. Hamada, Self-regerneration of a Pd-perovskite catalyst for automotive emissions control, Nature, 2002, **418**, 164.

23. J.M. Thomas and W.J. Thomas, Principles and practice of heterogeneous catalysis, VCH, NY, USA.

Fuel Cell Materials [24]

Fuel Cell:

Fuel cell is an electrical device that converts chemical energy into electrical energy by redox reactions at the electrodes. Fuel cell is currently attracting tremendous interest because its operation needs sustainable energy resources and its function leaves environmentally friendly non-toxic gases. The fuel cell consists of anode, cathode and electrolyte. At the anode, fuel such as H_2 is oxidized into proton and electron whereas at the cathode, oxygen is reduced into O^{2-}, which then combines with proton to yield H_2O as gas. The electrons released at the anode passes through external wire to cathode and thus, electricity is produced. Function of a fuel cell is described below (Fig. 20).

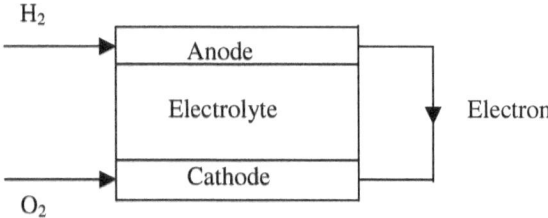

Fig. 20: Representation of a Fuel Cell.

The electrolyte is sintered material whereas the electrodes are porous materials. The main function of electrolyte is that it does not allow mixing of H_2 and O_2 before getting involved at the electrode reactions. The other property of electrolyte is not to conduct electrons from anode to cathode through electrolyte and it should be stable under oxidizing and reducing atmospheres. There are five fuel cells are available now-a-days and among them, proton exchange membrane fuel cell and solid oxide fuel cell are attractive because of their differences in functions. Therefore, we focus hereafter to the above mentioned fuel cells only.

I.   Proton Exchange Membrane Fuel Cell (PEM fuel cell)

Proton exchange membrane (PEM) fuel cell will play an important role in the next generation for the clean electricity production from the straight forward reaction between H_2 and O_2 at relatively low temperature, 70-80°C. H_2 gas is electrocatalytically oxidized to proton at the anode and at the cathode, O_2 from air is electrocatalytically reduced. Polymer membrane carries proton from anode to cathode which then combines with O^{2-} producing H_2O and heat. The electrons released during this process will flow from anode to cathode through an external wire and thus, electricity is produced.

II.   Solid Oxide Fuel Cell (SOFC)

In SOFC, solid oxide electrolyte, Y_2O_3 stabilized ZrO_2 transport oxide ion from cathode to anode where in fuel is oxidized into H_2O and CO_2. Its operating temperature is above 700°C. Other name of SOFC is ceramic fuel since the electrolyte employed is ceramics. The overall cell reaction occurring in SOFC is represented below.

$$H_2 + CO + O_2 \rightarrow H_2O + CO_2 + energy \tag{28}$$

The differences between PEM fuel cell and SOFC are summarized below in the table 3.

Table 3: Comparison of PEM fuel cell and SOFC

PEM Fuel Cell	SOFC
Operating Temperature is ~80°C	>700°C
Polymer electrolyte	Oxide ion conductor as an electrolyte
Electrolyte carries proton from anode to cathode	Electrolyte carries oxide ion from cathode to anode
H_2O is formed at the cathode	H_2O and CO_2 are produced at anode
Electrodes are Pt	Anode is cermet and cathode is oxides

Relevant Reference:

24.   R. Mark Ormerod, Solid oxide fuel cells, Chem. Soc. Rev., 2003, **32**, 17.

Thermoelectric Materials [25,26]

The title itself implies that this material deals with thermal and electrical energies. Except material synthesis, all the properties are physics oriented. As a chemist I try to explain it very briefly in a simplified manner. Thermoelectric material is a semiconductor that finds applications in thermoelectric devices. These devices, making use of property of semiconductor, is simple, free from noise pollution and no toxic or no green house gases evolve. Thermoelectric device converts thermal energy into electrical energy and as well as it cools one end when it is connected the other end to external input (Refrigeration).

Construction of Thermoelectric Device:

This device makes use of two different semiconductors, one is n-type and another is p-type semiconductors, connected in series. N-type semiconductor carries current by electron whereas p-type semiconductor carries current by hole. There are two applications found using these properties of semiconductors.

In a refrigeration, one end of two types of semiconductor is connected to external power. Charge carriers carry current as well as heat. Thus, external power attracts the charge carriers towards it. Because of movement of charge carriers towards external power, action of cooling takes place at the other end. This is due to carrying of heat by charge carriers as shown in the following figure (Fig.21).

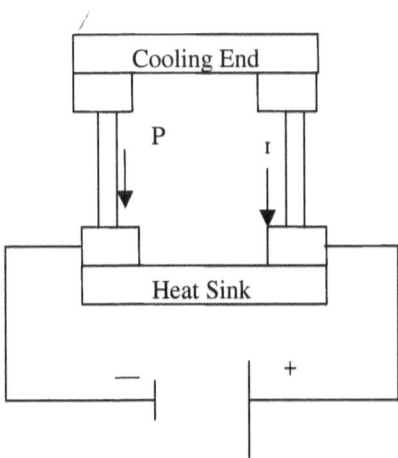

Fig.21: Action of refrigeration using thermoelectric materials.

When one end of two semiconductors is heated voltage develops across the two semiconductors. Thus, heat is converted into electricity. When one end is heated, charge carriers move towards heating end. The following figure represents the function of power generation using thermoelectric materials (Fig.22).

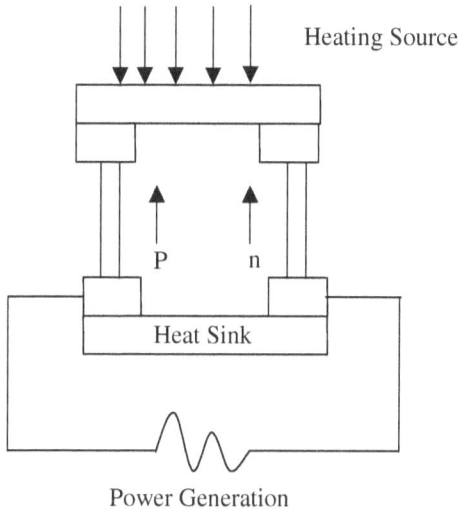

Fig.22: Function of power generation using thermoelectric materials.

The efficiency of thermoelectric materials is determined by ZT where T= temperature and Z is a figure of merit. Z depends upon the electrical and thermal properties of semiconductors. To be a better semiconductor, they should show the lowest thermal conductivity and the highest electrical conductivity. Semiconductors having heavier elements are used in these applications since the thermal conductivity is lower for them. Recently, oxide materials are also found to show better property in these applications. A thermoelectric refrigerator requires ZT = 3 at room temperature.

Relevant references:
25. B.C. Sales, Smaller is cooler, Science, 2002, **295**, 1248.
26. Ji-W Mooon, W-S Seo, H. Okabe, T. Okawa, K. Koumoto, Ca-doped RCoO$_3$ (R = Gd, Sm, Nd, Pr) as thermoelectric materials, J. Mater. Chem., 2000, **9**, 2007.

The Hydrogen Economy: The Forever Fuel, H$_2$

By 2020, the world will face an oil crisis. The majority of American automobile owners will suffer greatly from this crisis if new energy sources are not available. The first oil shortage in the 1970s and 1980s was economically and politically induced. This time, however, the crisis will be based on a real shortage of oil for fuel. Although optimist argue that new oil fields like exist 3280 feet or more below the surface of the oceans, the process of finding and obtaining it is very expensive and the technologies to do so are not fully developed. Therefore, it is highly desirable to avoid dependence of fossil fuel. Also, carbon-based oil produces carbon dioxide (CO_2) as a byproduct that will likely increase global temperature by 2.52 to 10.44^oF over the next one hundred years. These temperature increased will cause the polar ice caps to melt. Therefore, alternative to carbon based oil fuels must be researched. This could be achieved by using Hydrogen (H_2) as a fuel that produces only healthy water vapor when it burns, and does not heat up the air.

Chemical reactions could be explored to produce hydrogen gas. One such example is the violent reaction of reactive metal with acid. Thus, Zn reaction with HCl yields $ZnCl_2$ and H_2 gas. Similarly, Fe metal reacts with sulfuric acid leads to production of H_2 gas and $FeSO_4$. These reactions are represented below.

$$Zn + 2HCl \rightarrow ZnCl_2 + H_2 \text{ (gas)} \tag{29}$$
$$Fe + H_2SO_4 \rightarrow FeSO_4 + H_2 \text{ (gas)} \tag{30}$$

Another chemical reaction is that metals such as K, Na and Ca react vigorously with water producing enough heat to ignite the hydrogen. Example of Na reaction with water is represented by the following reaction.

$$2Na + 2H_2O \rightarrow 2NaOH + H_2 \text{ (gas)} \tag{31}$$

The formed NaOH in the above mentioned reaction further can be treated with Al or Si in presence of water to produce H_2 gas.

e.g. $$2Al + 2NaOH + 6H_2O \rightarrow 2NaAl(OH)_4 + 3H_2 \text{ (gas)} \tag{32}$$
$$2NaOH + Si + H_2O \rightarrow Na_2SiO_3 + 2H_2 \text{ (gas)} \tag{33}$$

Hydrides could be explored as a source for hydrogen. Calcium hydride or sodium borohydride is treated with water to produce hydrogen gas and the reactions involved in these processes are represented below.

$$CaH_2 + 2H_2O \rightarrow Ca(OH)_2 + 2H_2 \text{ (gas)} \tag{34}$$
$$NaBH_4 + 4H_2O \rightarrow NaB(OH)_4 + 4H_2 \text{ (gas)} \tag{35}$$

FEW NEW DIRECTIONS

Aqueous Combustion Synthesis of Oxide Semiconductors and Their Use for the Photosplitting of Water at Room Temperature

Background

Solar-splitting of water using semiconductors plays a vital role in photo catalysis field [27]. Using photon to split water into H_2 and O_2 IT is aimed at harness the solar energy because solar radiation reaching at the earth surface for 1 h is equivalent to fossil energy consumed by the world for 1 year. Therefore, research on photo catalytic breakdown of water is of world wide interest [28]. The reduction of water into H_2 and oxidation of water into O_2 using semiconductors are represented below.

$$2H_2O \xrightarrow[\text{Semiconductor}]{\text{Photon}} 2H_2 + O_2 \qquad (36)$$

The Fig. 23 represents energy profile diagram for equation 36. It is very clear that potential of product is greater than that of reactant and hence, it is a reversible reaction. In order to forward the reaction, products should be removed from the reactor continuously.

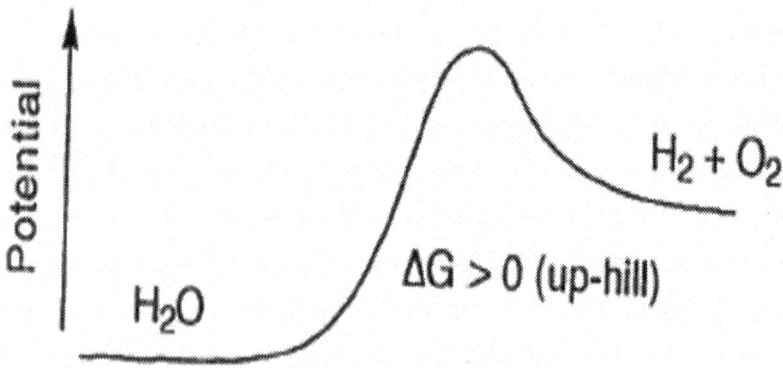

Fig.23: Energy profile for water splitting

The principle of photo splitting of water into hydrogen and oxygen using semiconductors with appropriate band gap and positions of valence and conductance bands is represented in Fig.12 (refer to page 26). The energy difference for the redox reaction of H_2O into H_2 and O_2 is 1.23 eV. Therefore, semiconductors with band gap greater than 1.23 eV are required for the solar-splitting of water into H_2 and O_2. Also, this photo evolution of gases from water depends upon the position of conduction and valence bands of semiconductors. Thus, the bottom level of the conduction band has to be more negative than the redox potential of H^+/H_2 (0 eV Vs NHE), while the top level of the valence band be more positive than the redox potential of O_2/H_2O (1.23 eV). It indicates that electron and hole are generated under photons with energy equal or greater than the band gap of semiconductor. Thus, electrons are promoted to the conductance band whereas holes remain in the valence band. These charge carriers are responsible for photo splitting of water. In a real situation, there are two main things occurring in a semiconductor particle and this is represented in the Fig. 24 below.

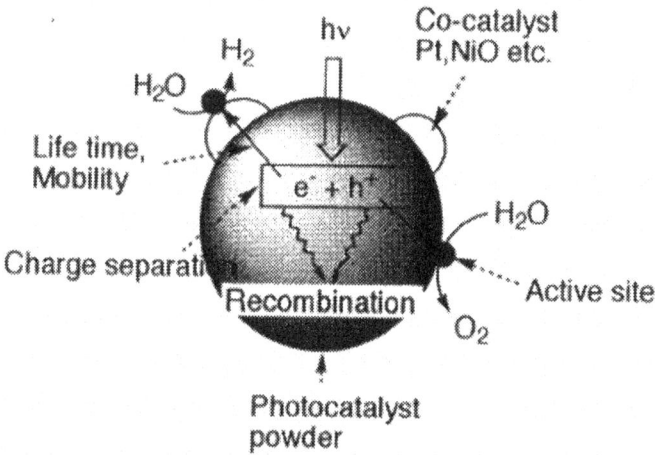

Fig. 24: Photosplitting of water using particle semiconductor

The desired thing is the charge separation of electron and hole and they migrate to the surface of the particle. In order to improve the catalytic activity, a co-catalyst such as Pt or NiO is coated at the surface of the particle. An undesirous process is recombination of charge carriers which lead to a lower yield for the reaction.

Among the semiconductors, TiO_2 is widely studied for photo splitting of water. TiO_2 semiconductor shows photo splitting of water into H_2 and O_2 under UV irradiation. Similarly, another system studied extensively is $K_4Nb_6O_{17}$ [29]. The success of $K_4Nb_6O_{17}$ semiconductor in the photo splitting of water is due the fact that hydrogen evolution and oxygen evolution take place in different interlayer space and hence, recombination of the gases is avoided. However, none of the layered compounds exhibited photo splitting of water into H_2 and O_2 under visible irradiation. On the other hand Pt/CdS is one of the photo catalysts that show evolution of H_2 gas using sacrificial reagent under visible irradiation, while WO_3 or $BiVO_4$ shows evolution of O_2 gas under the same conditions using different sacrificial reagent [30].

Aims

a) The forgotten research in solar-splitting of water is use of bulk/nano/transition metal ions doped semiconductors as photo catalysts even though there was report on complete splitting of water into O_2 and H_2 under visible irradiation using nickel doped $InTaO_3$ semiconducting oxide [31]. Therefore, now is the time to devote research activities toward bulk/nano semiconductors with or without doping transition metal ions (refer to table 4).

b) The other direction forgotten in this field is use of mixed oxide semiconductors. Of interest is $BiVO_4$ (mixed oxide semiconductor) which shows excellent and clean O_2 production in presence of sacrificial reagent. It is also interesting to notice that Bi_2O_3 is a semiconductor and V_2O_5 is another semiconductor. Combination of these two individual semiconductor leads to mixed oxide semiconductor. Therefore, I assume the following thumb rule (refer to table 5).

Simple oxide semiconductor + Simple oxide semiconductor

$$\text{Simple oxide semiconductor + Simple oxide semiconductor} \xrightarrow{\hspace{2cm}} \text{Mixed oxide semiconductor} \tag{37}$$

$$\text{E.g. ZnO + Fe}_2\text{O}_3 \xrightarrow{\hspace{2cm}} \text{ZnFe}_2\text{O}_4 \tag{38}$$

Similarly semiconductor rich mixed oxides could be potential semiconductors for photo splitting of water into H_2 and O_2.

c) Among the known various semiconductors, SnO_2 and WO_3 (Fig.25) have their lowest valence bands the most positive than that of redox potential of H_2O into O_2.

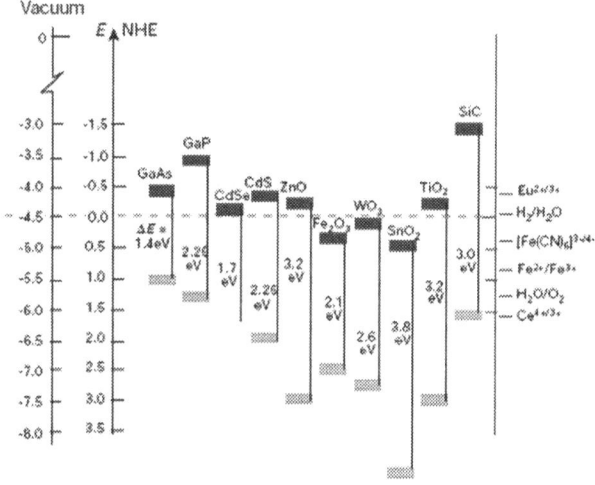

Fig.25 Energy levels of semiconductors.

Therefore, SnO_2 and WO_3 could be potential candidates for photo splitting of water into O_2 in the presence of sacrificial reagent. It is noted that these semiconductors can not produce H_2 from water under photon since their lower level of conduction band is less negative than that of redox potential of H_2O into H_2. Therefore, I would like to couple two complementary semiconductors as nano composites to photo-splitting of water into H_2 and O_2 without any sacrificial reagent. Thus, WO_3 or SnO_2 oxide will be incorporated into the layer of $H_4Nb_6O_{17}$ semiconductor, which is known to produce H_2 from water.

Approach

Now, I propose the following semiconductors for the solar-splitting of H_2O into H_2 and/or O_2 under visible irradiation.

a)   Table 4: Possible semiconductors with band gap in the visible range

Semiconductors	Band Gap, eV
CdO	2.00–2.20
SnO_2	3.4–3.60
In_2O_3	2.6–2.70
Cd_2SnO_4	2.0–2.20
$CdIn_2O_4$	2.7–3.00

b)   Table 5 is generated based on thumb rule as stated in the aim section.

Simple Semiconductor	Simple Semiconductor	Mixed Semiconductor or semiconductor rich oxide
ZnO	Fe_2O_3	$ZnFe_2O_4$
ZnO	SnO_2	Zn_2SnO_4
ZnO	In_2O_3	$Zn_2In_2O_5$
BaO (insulator)	TiO_2	$Ba_2Ti_9O_{20}$
BaO (insulator)	Nb_2O_5	$BaNb_2O_6$

III.A. Synthesis

Aqueous combustion method could be employed to synthesize the above mentioned compounds. This synthesis method is a wet-chemical method that occurs with evolution of heat and light. Thus, when rapidly heating aqueous solution of corresponding metal nitrates and fuel (glycine or urea or hydrazine based compounds) in the temperature range 350-500°C the solution boils, froths and catches fire. The in-situ temperature increases to about 1000°C. Such a high temperature is sufficient enough for the formation and crystallization of oxides.

Advantages of Aqueous Combustion Method

(i)   Molecular level homogeneity of metal ions is achieved.

(ii)   Very low ignition temperature (350°C-500°C) is sufficient for getting high "in situ" temperature (>1000°C).

(iii) Exothermicity is used for formation and crystallization of oxides

(iv)   The entire combustion reaction lasts for less than 5 min.

(v)   Combustion method is simple, fast and efficient.

Other synthetic methods employed to synthesize semiconductors include ion-exchange, intercalation and hydrothermal. Powder XRD, DRS, TGA-DTA, chemical analysis will be used to characterize semiconductors. Photo-fission of water into H_2 and O_2 will be evaluated by specially designed experimental set-up.

Relevant References

27. A. Kudo, J. Ceram. Soc. Japan, Development of photocatalyst materials for water splitting aiming at light energy conversion, **109**(6), 2001, S81-S88.

28. K. Sayama and H. Arakawa, Effect of carbonate salt addition on the photocatalytic decomposition of liquid water over Pt-TiO$_2$ catalyst, J. Chem. Soc., Faraday Trans., l**93**, 1997, 1647-54.

29. S. Ikeda, A. Tanaka, K. Shinohara, M. Hara, J.N. Kondo, K. Maruya and K. Domen, Effect of the particle size for photocatalytic decomposition of water on Ni-loaded K$_4$Nb$_6$O$_{17}$, Microporous Materials, **9**, 1997, 253-58.

30. H. Tanaka and M. Misono, Advances in designing perovskite catalysts, Current Opinion in Solid State and Materials Science, **5**, 2001, 381-387.

31. Z. Zou, J. Ye, K. Sayama and H. Arakawa, Direct splitting of water under visible light irradiation with an oxide semiconductor photocatalyst, Nature, **414**, 2001, 625.

Photosplitting of Water Using Nanocomposite Semiconductors with Open Framework Structure

Background

The search for new compounds in open framework materials (also called organically templated inorganic materials) stems from the fact that they exhibit wide variety of solid state structures [21] and also equally potential for industrial applications such as ion-exchange, molecular sieve and catalysis [32]. After the discovery of aluminium phosphate there has been extensive work to synthesize phosphates containing main group and transition metal ions. In addition to tetrahedral geometry, there are octahedral and five coordination geometries possible in the transition metal phosphates.

Since boron finds place in the periodic table just above aluminium there has been effort to synthesize open framework materials containing boron. The interest in boron based compounds arises due to excellent catalytic activity of bulk BPO$_4$. But, it is obvious that tetrahedral aluminium in aluminium silicate/phosphate can not easily be replaced by boron since boron prefers three fold coordination. However, there are minerals like datolite known to have tetrahedral boron in their compounds. Therefore, lot of efforts is needed to find out a suitable reaction conditions for obtaining boron in tetrahedral coordination. Photosplitting of water using semiconductor into hydrogen and oxygen is a clean fuel production and hence, bulk and layered semiconductors are widely used for the same

These materials are usually made under hydrothermal conditions containing alkali metal sodium silicate/phosphoric acid and sodium aluminate under basic conditions. Later the alkai metals have been replaced by organic amines. The organic amines act as template or structure directing agent for the formation of these materials. The organic diamines as cation occupy in channels (1-dimensional or 2-dimensional or 3-dimensional and exhibit hydrogen bonding with framework oxygen/hydroxyl group.

Aim

However, not explored semicoductor in the field of photosplitting of water into H_2 and O_2 is open framework materials with 1-dimensional or 2-dimensional or 3-dimensional channels. Therefore, research in this new and novel direction is removal of organic molecules from inorganic/organic hybrid materials without affecting open framework structure either by calcinations at low temperature <200°C or by photodecomposition at near temperature. And also to study photosplitting of water using these open framework materials. Nanocomposite semiconductors also will be designed for photosplitting of water.

Approach

(i) **Development of inorganic/organic hybrid materials with semiconducting property.**

Removal of organic diamines from hybrid materials at above 500°C by calcination leads to destruction of open framework of these materials. Thus, this calcination process converts open framework materials into dense materials and high surface area can not be retained in the dense materials. Therefore, it is aimed to expel organic molecules without affecting the open framework by photodecomposition. If open framework materials possess semoconducting property, it would be a judicious choice for ease removal of organic molecules by UV or visible irradiation [33]. Once, organic molecules are removed by photolysis the protonated form of open framework compound, having higher specific surface area, would be the final product. The protonated form of semiconductors with open framework is potential for photosplitting of water like $H_4Nb_6O_{17}$ semiconductor [29].

(ii) **Exploration of hydrazine as a template in the synthesis of open framework materials.**

So far, except one report [12], organic diamines are employed as a template in the synthesis of open framework materials. The one report is from my work with Prof. Slavi C Sevov at Notre Dame University and the template molecule employed was hydrazine i.e. carbon-free template molecules for the synthesis of cobalt phosphate. Hydrazine in cobalt phosphate donates one nitrogen for a coordination to cobalt ion. Therefore, removal of hydrazine without affecting the structure has been failed in cobalt phosphate system.

Usually, hydrazine based compounds have lower decomposition temperature compare to that of ammonia based compounds. Therefore, hydrazine based open framework materials may possess lower decomposition temperature and hence, protonated open framework materials could be obtained by calcinations at $<200^{\circ}C$.

The other way of removal of organic molecule from inorganic/organic hybrid materials is possibly by starting with mixed valence metal phosphate. Calcination of mixed valence metal phosphate converts lower valence metal ion into higher valence since organic cation is expelled out from the structure. Mixed valence metal phosphate could be synthesized by hydrothermal oxidation of transition metal powder with phosphoric acid in presence of organic diamine molecules.

(iii) **Nanocomposite Semiconductors for Photosplitting of water**

Charge carriers are formed in semiconductor under UV or Visible irradiation and these charge carriers are responsible for photosplitting of water. In order to have high rate of photosplitting of water, charge carriers have to be separated from their recombination. This can be achieved using nanocomposite semicondutors. Therefore, the research should be directed to synthesize nanocomposite semiconductors and to study photosplitting of water using nanocomposite semiconductors. The following flow chart gives subsequent reactions for obtaining nanocomposite semiconductors.

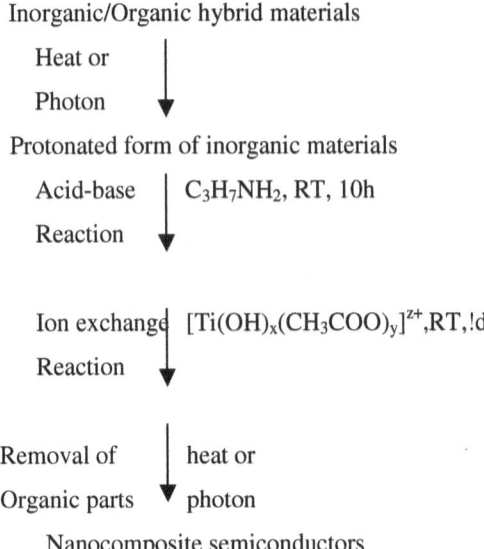

Inorganic/Organic hybrid materials

Heat or

Photon

Protonated form of inorganic materials

Acid-base $C_3H_7NH_2$, RT, 10h

Reaction

Ion exchange $[Ti(OH)_x(CH_3COO)_y]^{z+}$, RT, !d

Reaction

Removal of   heat or

Organic parts   photon

Nanocomposite semiconductors

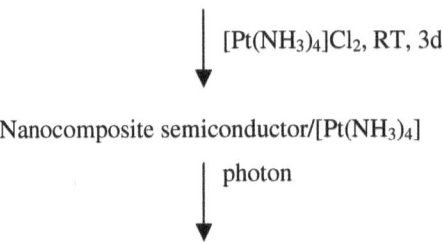

$[Pt(NH_3)_4]Cl_2$, RT, 3d

Nanocomposite semiconductor/$[Pt(NH_3)_4]$

photon

Nanocomposite semiconductor/Pt

Table 6 summarizes selected semiconductors that find application as photocatalyst.

Table 6: Selected Semiconductors [34,35]

Compound before decomposition	Photodecomposable moity	Expected Product
$M_4In_{16}S_{33}.(H_2O)_{20}.$ $(C_{10}H_{28}N_4)_{2.5}$ Where M = Cd and Zn	$C_{10}H_{28}N_4$	$M_4In_{16}S_{33}.(H_2O).(H^+)_{10}$ Where M = Cd and Zn
$[(CH_3CH_2)_2NH_2]_6In_{10}S_{18}$	$[(CH_3CH2)_2NH]$	$In_{10}S_{18}(H^+)_6$
$[C_6H_{16}N]_4In_4S_{10}H_4$	$C_6H_{16}N$	$In_4S_{10}H_4.(H^+)_4$
$[C_{13}H_{14}N_2]_4.In_9S_{17}$	$C_{13}H_4N_2$	$In_9S_{17}.(H^+)_7$

Relevant References:

32. M. Hartmann and L. Kevan, Transition-metal ions in aluminophosphate and silicoaluminophosphate molecular sieves: Location, Interaction with adsorbates and catalytic properties, Chem. Rev., 1999, **99**, 635.

33. H. Li, J. Kim, T.L. Groy, M. O'Keeffe and O.M. Yaghi, $20\overset{\circ}{A}$ $Cd_4In_{16}S_{35}{}^{10-}$ supertetrahedral T4 clusters as building units in decorated cristobalite frameworks, J. Am.Chem.Soc., 2001, **123**, 4867.

34. C. Wang, Y. Li, X. Bu, N. Zheng, O. Zivkovic, C.S. Yang and P. Feng, Three-dimensional superlattices built from $(M_4In_{16}S_{33})^{10-}$ (M = Mn, Co, Zn, Cd) supertetrahedral clusters, J.Am.Chem.Soc., 2001, **123**, 11506.

35. C.L. Cahill and J.b. Parise, On the formation of framework indium sulfides, J.Chem.Soc., Dalton Trans., 2000, 1475.

Design, Synthesis, and Photocatalysis of Semiconductors with Open Framework Structure

Thrust in finding a suitable semiconductor with open framework structure for photodetoxification of organic contaminants is the world wide interest. This is due to large surface area expected for open framework structures. These proposed semiconducting compounds resemble microporous zeolites.

Conventional semiconductor is bulk TiO_2. TiO_2 exhibits photodetoxification of organic contaminants under UV irradiation. Therefore, it is highly desirable to develop microporous compound consisting of TiO_2. The goal of proposed research is to design, synthesis and study of microporous titania and other semiconductors.

It is well known that titanium metal powder can be easily oxidized hydrothermally to titania by hydrogen peroxide and it is represented by the following equation (39).

$$Ti + 3H_2O_2 + 2OH^- \text{--------}\rightarrow TiO_4{}^{4-} + 4H_2O \tag{39}$$

<u>Proposed Chemical Reactions:</u>

$$\text{Hydrothermal}$$
$$Ti + 3H_2O_2 + Diamine \text{--------}\rightarrow Product \tag{40}$$

$$\text{Metal hydroxide} + H_2O_2 + \text{Diamine} \xrightarrow{\text{Hydrothermal}} \text{Product} \qquad (41)$$

$$\text{Metal powder} + \text{Oxidizer} + \text{Diamine} \xrightarrow{\text{Hydrothermal}} \text{Product} \qquad (42)$$

Where Metal = Zn, Cd, Ti, Nb, W, Bi, In, Sn and V.
Oxidizer = KI-I, $KMnO_4$, Ammonium thiosulfate.
Diamine = Ethylene diamine or 1,12-dodecanediamine or 1,8-octanediamine or hexanediamine.

Experimental:

Hydrothermal technique is a feasible method for the synthesis of microporous semiconductors. It involves heating corresponding metal powder or metal hydroxides with oxidizer in presence of organic diamines as a template in the temperature range 100–200°C under autogeneous pressure. The hydrothermal technique is carried out using autoclave.

Open Framework Materials

Open framework materials are proven to be versatile compounds in the subjects of inorganic and materials chemistry because of their potential applications such as molecular sieves, catalysts and ion-exchangers. Now-a-days formation of open framework materials is extended to transition and rare-earth metals and therefore, new research areas such as magnetism and luminescence are growing up. The open framework materials are usually obtained by two ways. One of them is the use of organic diamines. The organic diamines act as template to direct the formation of open framework structures, free from formation of dense compounds and equally they satisfy the charge compensation. The templates form extensive hydrogen bonding with framework oxygen or hydroxides group. Another way is the use of organic ligands to get open framework compounds. In some cases, both template as well as covalent or coordinate bonding are observed.

The research in this field is the exploration of hydrazine based compounds since their dual behavior such as template and ligand. The lone-pair of each nitrogen is readily available for bonding with metal ions and also it is possible for protonation of both the nitrogen under acidic condition to act as templates or protonation at one-end for template and bonding of another nitrogen with element. Thus, positive ligand in the formation of open framework materials is exciting one.

Out of several compounds, tetraformyl trisazine (TFTA) is a promosing compound for the study of open framework structure. In addition to anionic open framework structures, cationic structure of boat or chair form of TFTA will be evolved in this area.

Synthesis of TFTA:

$$3NH_2\text{-}NH_2 + 4\ CH_2O\ \text{(formaldehyde)} \dashrightarrow C_4H_{14}N_6O\ \text{(TFTA)} \quad (43)$$

Synthesis of open framework materials:

$$\text{Hydrothermal}$$
$$\text{Metal salt} + \text{TFTA} \dashrightarrow \text{Metal-TFTA framework} \quad\quad (44)$$

Structure of TFTA:

978-0-595-39374-9
0-595-39374-8

www.ingramcontent.com/pod-product-compliance
Lightning Source LLC
Chambersburg PA
CBHW021025180526
45163CB00005B/2113